FARM TALK:

A SERIES OF

ARTICLES IN THE COLLOQUIAL STYLE,

ILLUSTRATING

VARIOUS COMMON FARM TOPICS.

BY

GEO. E. BRACKETT,

BELFAST, MAINE.

BOSTON:

LEE AND SHEPARD.

1868.

STEREOTYPED AT THE BOSTON STEREOTYPE FOUNDRY,
No. 19 Spring Lane.

THIS LITTLE VOLUME

IS

Respectfully Dedicated

TO

MY FRIEND AND FELLOW-LABORER

S. L. BOARDMAN.

CONTENTS.

CHAPTER		PAGE
I.	GUESS FARMING.	9
II.	PEDIGREE CORN.	15
III.	ABOUT HAYING.	20
IV.	FANCY FARMERS.	26
V.	WHEN TO SELL PRODUCE.	30
VI.	BUTTER MAKING.	36
VII.	GETTING READY FOR THE CATTLE SHOW.	41
VIII.	AGRICULTURAL COLLEGES.	47
IX.	APPLE TREES AND INSECTS.	54
X.	MIDDLE-MEN.	61
XI.	TAKING THE PAPERS.	66
XII.	THE 'OLOGIES.	70
XIII.	AN EVENING'S CHAT.	75
XIV.	PLANTING FOR POSTERITY.	80
XV.	PARASITIC PLANTS.	86
XVI.	ROAD-MAKING AND BREAKING.	92
XVII.	IN THE BARN.	97
XVIII.	HOW TREES GROW.	104
XIX.	PIGS AND POULTRY.	109
XX.	FARM FENCES.	115
XXI.	OUT IN THE FIELDS.	122

FARM TALK.

I.

GUESS FARMING.

WE farmers are a stubborn class to learn. We do not accept facts without a good deal of proof and persuasion, and in too many cases work out our own injury through a fear of being too easily deceived. This is a progressive age, and those who allow themselves to fall in the rear in the march of improvement must be content to occupy second-rate positions, and be satisfied with small pecuniary rewards.

It is somewhat surprising, that, notwithstanding the improvements in nearly every department of farming, so little has been effected towards inducing farmers to perform their business operations in a more systematic manner, and keep a regular record and account of their farming and business, generally and specifically. But for one farmer who practises such a method, ninety-nine keep all their accounts "in their head," as it is termed, and consequently are properly called *guess farmers*, for they never *know* anything, only *guess* it is so and so. They don't know whether this, that, or the other crop pays best; whether they can afford to sell their stock or produce at such and such a price or not. They can't tell whether it is for their interest to continue a certain course of husbandry, use such a fertilizer, cultivate a soil in such a manner, nor even at the end of the year are they sure whether their names should be recorded on the profit or loss side of the ledger. Having kept no account of their doings, they are almost wholly in the dark. They can only "*guess* it's about so."

These ideas were forcibly brought to my mind last evening, as my neighbor Smith came in while I was writing in my farm record, and, noticing my occupation, said, with a sly twinkle in his eye, —

"What you doin', — book-keepin'?"

"Yes, a little. Squaring up my accounts for this year. You know to-morrow is the first day of January, and I always like to commence the new year with a clean sheet, and to know how I stand with the world in general, and my fellow-men in particular. Besides, I want to know whether it has paid me to farm it this year. I suppose you keep farm accounts, don't you?"

"Me? No. It's too much bother. I can keep my 'counts in my head."

"Don't you think it's better to have something you can rely upon? You know we are all apt to be forgetful."

"It's well enough for store-keepers, and sich, to keep 'counts; but I don't see no need of a farmer spendin' time doin' it."

"Isn't it as much for our interests to look after and keep posted in our own business, as it is for the merchant to attend to his?"

"Well, yes, I s'pose so. But what is the use of a feller's writin' down everything about what he does?"

"It *pays* to do it, Smith."

"I don't see how."

"Well, look here. How was your corn crop this year?"

"Pretty fair; though the frost hurt it some."

"Shall you plant some more next year?"

"Sartin. I allus plant a piece o' corn."

"What for?"

"What for! Why, because—because I allus have."

"Yes, I see. Now, Smith, how much did that field of corn pay you?"

"Pay me! Well, I guess about ——"

"That isn't it. I don't care what you *guess;* do you *know?*"

"Well, not exactly; but I cal'late ——"

"Never mind. But really, now, do you know whether you made or lost money in raising that field of corn? There's the rub."

"Dunno as I do, for sartin."

"Now, see here, Smith. Here is my 'Corn Field Record.' I have written down everything connected with it, and to-day I have summed it all up, and *know* all about it; there's not a particle of guess work. Here is the size of the field; kind of soil; when and how many times it was ploughed, harrowed, and furrowed;

and also the amount and kind of manure ; how it was applied, and what and how much top-dressing I used. Then there is the time of planting; preparation, and kind of seed; how far apart it was planted in the rows and hills, and how many stalks in the hill; when and how long after planting it came up; manner of cultivating, time of harvesting, &c. And here I have got, in exact figures, the cost, value, and price of everything; planting, hoeing, and harvesting; the value of the corn, beans, pumpkins, and fodder; the value of the manure, rent of land, manure left in the soil, and every item set down in full, so that I know exactly what it cost me to raise that piece of corn, and thus whether it pays to continue to grow corn under such circumstances."

This is only an example. All other farm operations may be recorded in somewhat the same manner, more simple, if you please, and then there would be an end of the continual *guessing* and *thinking*, and ignorance in regard to those things we should *know* about. Of course keeping such records occupy some time, but how can it be spent to more advantage than in thus obtaining a better knowledge of our profession, and the workings of that special branch in which we may be engaged? Brother farmers, think of this matter, and act upon it. Don't be careless and stubborn, and persistently continue to drive on in the old ruts, simply because you have been accustomed to so doing, or because your fathers have done so before you; but accept a change whenever it is for your advantage to do so, if not for your own sakes, for the good of your boys and posterity.

II.

PEDIGREE CORN.

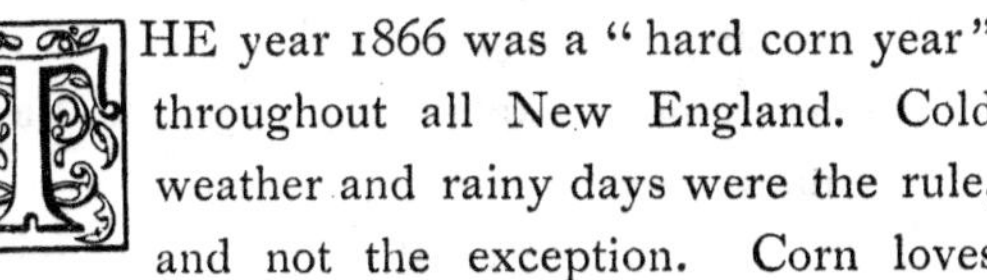

HE year 1866 was a "hard corn year" throughout all New England. Cold weather and rainy days were the rule, and not the exception. Corn loves warmth and dryness, and pines in a cool, damp atmosphere. The two preceding years were the warmest and dryest on record, consequently were what farmers term "great corn years." Corn grew and matured planted most anywhere. But this year it was different. There was a fair yield, but it was very backward, and many fields were frost-bitten before the kernel was fairly "glazed." An unusually wet and rainy September was a great drawback to its ripening. Notwithstanding these failings on the part of the season, my neighbor Johnson raised a good crop of corn, and some specimen ears, which he exhibited at the County Fair, attracted a good deal of attention.

"That's good corn for this year," said a farmer, pointing to Johnson's fine-looking trace.

"Yes," said another; "wonder how he was so lucky as to get such a good crop."

"Guess he must have used a big pile of manure, and had some warm soil too, not to speak of top-dressing," said a third.

Here Johnson happened along, and they began questioning him.

"What kind of land did this corn grow on?"

"Rocky upland, well drained."

"What did you manure it with?"

"Hog manure, mixed with muck, and a handful of plaster and ashes to each hill after the corn broke through the ground."

"Ground ploughed in the fall?"

"No; broke up in the spring, harrowed well, and planted on the sod."

"Like it so better than old ground?"

"Yes; the turf heating and rotting, heats the soil just when the young corn-roots need the warmth most; and, besides, the ground is not so weedy, requiring less labor in cultivating."

"Dung it in the hill?"

"Yes; one large shovelful to a hill."

"Like hog dung better than old manure?"

"Certainly. It's much better for corn."

"What kind of seed was yourn?"

"Well, we call it the Dutton corn, generally; but I don't know what is the exact name of it. However, I call this my 'Pedigree' corn; and I consider the principal reason why I had such good luck this year was because I took pains with my seed."

"What did you do to it?"

"Well, gentlemen, I'll tell you all about it, and why I call this my Pedigree corn. For several years it has been my practice to select the best ears of corn to be found in my field, and preserve them for next year's seed. By so doing, I think the quality and productiveness of my corn have been gradually improving year after year, until now I can see that it is much superior to that first grown."

"Does the cob have a red peth?"

"Sometimes red and sometimes white."

"I like red peth corn best."

"I never saw any difference in the quality of the corn, whether the cob had a red or white pith, and am of the opinion that both may be

2

found on the same stalk; — but I will go on about my seed corn. After the corn is thoroughly ripened, I make my selection. I take only those ears which are rowed perfectly, fully ripe, and golden yellow, with the end tipped or covered with kernels, and only those which grow on a stalk producing two or more sound ears. These ears I carefully trace or braid together by the husks left on them for this purpose, and hang them away in a dry place, where they will not be exposed to extreme cold, and beyond the reach of vermin. This I do every fall, and the next spring I have a superior article of seed corn, which I have christened my Pedigree corn."

"Well, I don't know; sometimes I've planted the best ears I could pick out of the lot, and sometimes only the tip ends and common ears, and I didn't see much difference in the crop."

"Perhaps not. I do not say that the improvement will be discernible in one year; perhaps not in two; but I do say that such a practice, persistingly followed for a term of years, will be productive of good results. It is a great law of Nature that like produces like, to a great extent, and the principle is as correct in reference to seeds and plants, as in the case of the higher orders of animal life. This is a fact which we, as farmers, do not sufficiently consider. We are too apt to be willing to dispose of the best of our flocks, herds, or crops, and use the poorest for planting, sowing, or breeding. Let us remember that the best we can raise or obtain is none too good for our own use, and that growing or breeding from inferior seed or animals always has a tendency to deteriorate, or run out the quality of the production, and *vice versa.*"

III.

ABOUT HAYING.

OME, boys, hitch up the mower, and we'll cut the ten-acre lot to-day. Here it is the 10th of July, and no grass cut down yet. I don't believe in feeding all straw to the stock next winter, but I do believe in feeding dried grass, and if that field remains much longer, there will be nothing but dead stalks to cut. Early cut hay is far more valuable to feed out than that which is not cut until it is ripe."

At this juncture, neighbor Smith, who is looking on, and who usually times his haying operations by mine, because, he says, I have a "b'rometer," takes a fresh quid, and inquires,—

"Ain't ye beginnin' a little airly?"

"No. If the season hadn't been so backward I should have commenced a week ago; but, as it is, I hope to get my grass cut in time to be palatable to the animals next winter."

"But don't you think it spends better not to cut it till it's about ripe?"

"Perhaps it will spend better if cut then; but the question is, when to cut it so that the hay will make the best quality of food; and, from experience, I have found that period to be when the plant is just opening its blossoms. Besides, I have submitted the question to a competent jury, which was my stock last winter, and they unanimously agreed with me, and if they are not judges of good hay, who is?"

"But them scientific fellers say it's got the most goodness in it when it's ripe."

"Yes, I know they say various wise things, which, unfortunately, do not always prove to be practicable. Now, I don't pretend to know what is the amount of gluten, starch, sugar, or other elements contained in the grass which I cut so early, but I do know that my stock eat it up cleaner, like it better, and keep in better condition when fed on it, than when kept on hay which remained in the field until it was ripe before it was mown; and that is all I am particular about knowing of the matter, from a practical point of view."

"Well, I always kinder thought they liked it better, but I didn't know."

"Certainly they do. Stock will eat early cut hay, even if it is of second or third quality, and leave but few orts. Grass in which there are a good many weeds and brakes should always be cut early. The English grasses will depreciate the least by leaving until nearly ripe. Fodder which, if allowed to remain until it has seeded, is worthless, if cut quite early, makes very good feed for sheep and young stock. What old lady nurse would think of letting her herbs remain until dead ripe before cutting them? No, she gathers them while in blossom, 'so as to get all the good of them,' she says. It's just so with the plants we use for fodder for our stock — cut them while in blossom, and we get all the good there is in them."

"It don't look like being a good hay day, it's so cloudy."

"Clear days are not always the best hay days."

"What kind of weather do you think is best?"

"I consider a cloudy day, with a fresh, warm, south-west wind, preferable. A clear day and hot sun is not essential to haymaking. There is a great error, which too many of our farmers fall into, — they 'make' their hay altogether too much. To make good hay, grass requires to be thoroughly dried, not scorched and burned. No doubt but the very best method of curing hay is to do it under cover; but this is not practicable to any considerable extent. The best herdsgrass hay I ever saw was mowed after the dew was off, put into the barn green, spread out thin, and thus made under cover, without allowing the sun to fall on it."

"How about clover?"

"Clover should be mown when dry, wilted in the sun, and then, if possible, made in the cock by sweating. It can then be handled without breaking off and losing the leaves, which are the most valuable portion of the plant. Yes, sir, it's my opinion we 'make' our hay altogether too much."

"'Twouldn't make much difference when you cut your hay, if you was goin' to press it up and sell it, would it?"

"Perhaps not; but it's only in very rare and particular cases a farmer can afford to sell his hay."

"Does your b'rometer say fair weather to-day?"

"Yes; the mercurial column is pretty high, and has been so for some days."

"Is that always a sure sign of good weather?"

"Generally so; but I find a barometer is not always to be depended upon. It requires a good deal of practice and observation to enable one to judge of the weather from its changes."

"It falls for storms, don't it?"

"Most always. But it is not always a true prophet. It will sometimes fall and rise without any corresponding changes in the weather. Yet there are many times, as in the case of sudden squalls and showers, when it is sure to give the alarm."

"Does it pay to have one?"

"Yes, I think it does. I remember one instance last haying season. We had experienced a three days' fog storm, with the wind still fresh to the east. In the morning I noticed the mercury in my barometer was rising fast, notwithstanding it was still thick and foggy, and I knew therefore we should have fair weather in twelve hours or less; so we set our men to mowing, and worked nearly a day, while our neighbors' crews were lying still, fearing the weather. It is also handy about foretelling thunder storms when there is hay out. But, though I have observed and studied one almost daily for many years, yet even now I would not pretend to be a wiser weather prophet than many of our old experienced farmers."

IV.

FANCY FARMERS.

HAT do you think of paying five hundred dollars for a ram?"

"Why, I think it's a pretty big price."

"Big price! It's a regular swindle."

"Don't you think such prices are paid?"

"No, I don't. I always calculated the reports of such great prices were humbugs, got up by the fancy farmers, who, I think, do more harm than good in any neighborhood. Their fast horses, thousand-dollar sheep, and big roosters, are altogether too 'fast' for us common farmers."

"Well, neighbor, I partly agree with you. No doubt there have been many falsely reported sales, for purposes well understood; but, on the other hand, a large number of rams have been sold in this country during the last five years at prices ranging from five hundred to upwards of a thousand dollars each."

"But were they worth it?"

"Ah, that's another question, and one which, in my opinion, must be answered in the negative. It is an old saying, that a thing is worth just what it will bring; but here is an exception to the general rule. A sheep-fever, or mania, rages throughout a portion of the country, and rams of the most fashionable breeds command prices far above their real value."

"But, neighbor, when you say that fancy farmers cause more harm than good to the locality in which they live, I must dissent from your position. As the term is usually understood, fancy farmers are those who, possessed of capital, expend it lavishly upon their farms in various directions, according as their tastes or prejudices lead them, and not always in an economical, or even, perhaps, sensible manner. They buy fast horses, blooded stock, fancy sheep and poultry, artificial manures, improved implements, search for new seeds and fruits, and are always the first to take hold of any 'newfangled notion' that starts up. They are usually men who have made their money by some other profession than farming, and consequently are not deeply versed in its mysteries, by experience. They are theorists,—fanatics, if you will,—and they enter upon new schemes with more impetuosity than wisdom. In a word, they 'run things into the ground,' and in nine cases out of ten, their experiments leave them poorer in purse, but richer in experience.

"Now, we insist that such farmers—amateurs, gentlemen farmers, fancy farmers, or whatever you see fit to term them—are acquisitions to any farming community. They are the pioneers in improvement—the extremists, who lead the way in the path of progression. By their operations a whole neighborhood may profit. They conduct an experiment;—if it is successful, others may profit by it without running the risks; if it is unsuccessful, we need not attempt, and have lost nothing. So no farmer, who keeps his eyes and his ears open, but may profit by the operations of his fancy farmer neighbor. He buys blooded stock, and pays what you consider an outrageous price. Very well; he pays his own money, and if he is satisfied, we ought to be. He keeps this stock, and if of value, in time it becomes mixed through all the surrounding herds, and we are the recipients of the gain, without being obliged to join in the expenditure. So of fertilizers and implements. He sees a new kind of machine or manure advertised, obtains it, and makes a trial of it. What is the result? If it is a success and pays, you can do the same; but if it is a failure, you are none the poorer, rather richer in experience obtained through the use of another's capital.

"Thus it is; and all this outcry against 'fancy farmers' is just so much spleen, prejudice, and a mistaken idea in regard to the matter. To be sure it may make us feel a little envious to see our new neighbor, fresh from some occupation in which he has acquired money, erecting nice buildings, purchasing fine stock, improving his fields, driving a stylish team, and spending his money generously, perhaps to us foolishly; but such is not the right spirit. We should strive to banish all such unworthy feelings, and, making the best of what we have, endeavor to turn all his experiences to our own advantage, and learn from his failures as well as his successes."

V.

WHEN TO SELL PRODUCE.

OTATOES have got up to fifty-five cents a bushel: do you think a feller had better sell now, or hold on to 'em a while longer?"

"Sell when it will *pay* to sell."

There is no question so often asked us during the fall and winter as, "When do you think I had better sell" this, that, or the other crop? and, "Do you think" such and such a thing "will be higher or lower?" To the former, our invariable answer is, Sell when it pays for you to do so; and to the latter, We cannot tell you, for in these days of changes and fluctuations, no man can tell what a day may bring forth as regards the produce markets. Of course there are certain general facts which have a bearing upon the matter, but these every farmer understands who is posted, as he ought to be, and may be.

"But how can I tell when 'twill pay?"

"There's the rub, Smith. You don't know when, or at what price it will pay for you to sell your potatoes, because you don't know what they cost. Isn't that a fact?"

"Wall, I don't know as I do know what they cost, adzactly; you see I never reckoned clust."

There, farmers, is the whole truth in a nutshell. You don't know at what price you can afford to sell your produce, because you don't know what it cost you to raise it. Isn't it so? How many of you can tell what your crops cost you in the aggregate, or by the ton, bushel, or pound, as the case may be; or can tell whether you are making or losing, whatever the market price may be? We think we are safe in saying, not two in a hundred. You think, guess, calculate, about so much, but you don't *know*. Wheat may be forty cents or a dollar and a quarter, and potatoes fifty or seventy-five cents a bushel, but you don't know whether it will pay for you to sell yours or not, because you can't tell what they cost. And so you "go it blind." If you hit a price that pays, you consider it a stroke of good luck; and if you lose, you "grin and bear it." But to return to Smith.

"What would you call paying?"

"When the amount received equals the cost of production, with a fair percentage on the capital employed."

This is what we consider sound advice, and the true policy, it being a safe course under all circumstances. If the market price is not sufficient to make good the cost of production, you are fully justified in holding on to your produce, as a general rule; but if it will return that, and good interest on the money in addition, you are on the safe side to dispose of it. But Smith says, —

"You can't keep potatoes always, you know."

"Very true. Those products which are of a perishable nature must be disposed of before another growing season, even at a sacrifice. Potatoes, and all kinds of vegetables and fruits, must be sold in their seasons, else they are a loss on the producer's hands."

"'Tain't always good weather to haul stuff to market, neither."

"That fact must also be taken into consideration. Late autumn and the winter season is the time the farmer can best attend to his marketing, and it will not pay to haul any distance over bad roads, or in the busy season of the year, for a few cents more on a bushel or pound. Could you afford to leave off in planting time to haul your potatoes to market, even if they were ten cents a bushel higher than in the winter?"

"No; not for twenty cents more."

"Of course not."

"I don't know, after all, but the best way to market your crops is to sell it on four legs."

"What do you mean by that?"

"Why, eat it up on the farm; feed it out to stock, and sell the animals."

That course is certainly preferable for those farmers who live at a distance from the markets, particularly the hay and grain; but large quantities of potatoes are not convenient to feed out, and I much doubt the expediency of so doing if they command fifty cents a bushel, when corn can be bought for a dollar.

"Do you cal'late a bushel of corn is worth as much as two of potatoes for stock?"

"Pretty near. Potatoes are first rate for an occasional feed, particularly with hay or dry fodder; but corn meal is the richest and best of anything in the shape of grain as food for stock."

"Wall, I don't s'pose a feller can tell what his crops cost him, unless he keeps some kind of reckonin' or 'counts, can he?"

"That's it, exactly, Smith. We farmers are altogether too slack and negligent in our business matters. What would you think of a merchant who knew nothing of the cost of his articles, and sold just as it happened? Of course he would soon fail, and he would deserve to. We can all see the impropriety of such a course. Then, why not apply it to our own business, our own profession? If anything, it is more complicated than the merchant's, and therefore requires more attention, and a stricter approach to method and system. Let us look before we leap. Forewarned is forearmed; or, as the modern adage has it, be sure you're right, then go ahead. But we are not sure we are right, and never can be, until the present slip-shod method of farming is done away with, and a system is inaugurated which is more in accordance with the needs of those who obtain a living by the cultivation of the soil."

VI.

BUTTER MAKING.

HILE making a call on my neighbor, Squire Brown, the other evening, the conversation turned on butter making and the dairy. The Squire keeps a small but fine dairy, consisting of a half dozen cows or so, and his wife is well known as a superior butter maker and dairy woman. I asked Mrs. Brown if she had any secret method of operation for making such nice butter at any and all seasons of the year.

"O, no, indeed! Only the method I've followed during these last twenty years."

"But your butter brings an extra price in market, and is always so hard, yellow, and nice-looking, whether it is winter or summer. How do you save your cream?"

"I skim the previous day's milk every morning, keep it in a stone jar, in a cool place, and churn as often as necessary, — say, three or four times a week, according to the season."

"What kind of a churn do you use?"

"The old-fashioned 'dasher.' I've tried several of the 'patents,' but none of them suit me. It's a great deal of labor to keep them clean."

"How do you 'lay down' your butter?"

"I make most of it into balls, to suit my customers, who like it better so for table use The rest I pack into stone jars, cover it with a layer of salt, and keep in the cellar until wanted for use or the market."

"What quantity of salt do you use to the pound?"

"A very little over an ounce to a pound. But some of my customers like it salted lighter. I am always careful to have the very best fine salt that can be found. Another, and I think one of the principal things in making good butter, is, to keep everything you use perfectly sweet and clean; which requires a strict personal attention, as well as plenty of scalding water."

"How do you operate with your cream in cold weather?"

"Warm the churn with scalding water, and set the cream near the fire, if possible, until warm enough. Or warm water may be added to the cream before churning, until it is of the right temperature."

At this point, the Squire, who had been quietly listening to the conversation, broke in with, "That's all very true, but you haven't given me any credit. I don't believe you can make bricks without sand, neither do I believe you can make good butter without good milk; and I know you can't have rich milk and cream without good cows, well fed."

"That's my opinion, exactly, Squire; and I've no doubt but Mrs. Brown's success in butter making is due as much to your out-of-door management as to her excellent method of operation. Are your cows any particular breed, Squire?"

"Well, no. The older ones are natives, and the younger ones are part Ayrshire blood. When I buy or raise a good cow, it don't make any odds to me what breed or blood she is. The proof of a cow is in the milk-pail, though I s'pose blooded stock is best to breed from."

"At what age do you have your heifers 'come in'?"

"The spring they are two-year olds, always. I don't believe in waiting till they are three-year olds, as some do; it's a dead loss of one year. I've tried it often enough to convince me. They make just as good cows to come in at two as to wait longer. To be sure the heifer gets more growth on her if she don't calve till she's three years old, but I never believed in monstrous cows. Of course I'm speaking of heifers which are well kept, and are thrifty, and large of their age. Only early calves, such as come in February or March, are suitable to raise. Then they can be weaned on hay, and, with a good pasture, will come to the barn in as good condition to winter as common yearlings."

"What is your general feed for your cows during the winter, or while they are at the barn?"

"Plenty of good English hay, with one feed a day of either roots or meal. Those that are in milk get a little extra, for it pays to do it. I give them as much clean water as they want twice a day, and let them run in the yard sunny weather. Get good cows, and give them generous food; keep them warm and clean, and use them kindly, and there is no branch of farming that is more pleasant, or pays better."

VII.

GETTING READY FOR THE CATTLE SHOW.

UR County Agricultural Society's annual fairs and cattle shows, or exhibitions, are very powerful influences towards elevating and bettering the condition of the common farmer, and advancing the cause of agriculture, if they are properly conducted. But that's what's the matter. Too many of them are made the vehicle for setting forth the peculiar ideas of certain men, and too many more are conducted in a manner not in accordance with the intentions of the act incorporating them. One makes flowers a speciality, another favors sheep, and a good many "run all to horse,"—horse walking, trotting, and racing, with all the usual concomitants,—intemperance, gambling, and immoralities. But, we are glad to say, that horse-racing, in connection with our agricultural exhibitions, is every year growing less frequent, and the leading

farmers and agricultural papers in the country are taking a firm and decided stand against it.

This matter of cattle shows was brought to my mind yesterday while over to neighbor Smith's. He is preparing for the cattle show; in fact, has been preparing for the last six months. He has been getting ready at the rate of a little less than a dollar a day. The particular line in which Smith will exhibit this year is in bulls. He is raising a bull calf, which, he says, he "is bound shall take the shine off" anything in our county, and it is my private opinion he will get the premium. You see, last February, one of Smith's cows dropped a large bull calf, and he immediately conceived the idea of making a premium animal out of him. So he has had the milk of the dam, and also most of the milk of another cow, all summer, and as a consequence has grown rapidly, being now about the size of one of my yearlings. Smith has taken the very best care of him, feeding plenty of grass, and what meal he would eat, and now he is a miracle of size and fatness.

"What do you think of him?" said Smith to me, looking admiringly at his bovine pet.

"He's large, certainly, and sleek too."

"Won't he make a splendid bull!"

"Well, he looks like it; but you can't always tell when they are so young."

That was as fully as I dared indorse Smith's opinion. When I remembered that the calf's sire was a "scrub" bull, and the dam only a common native, I knew that nine chances to one the calf would make an inferior stock animal. This is why we need blooded or thoroughbred animals for stock getting or breeding,—they almost invariably mark their "get" with their own valuable peculiarities and characteristics; whereas, if a native-bred bull be used, be he never so well-formed and fine an animal, his offspring are likely to "cry back" to his scrub parentage. These are facts, well grounded, and should be duly heeded by every stock raiser, be his speciality cattle, horses, sheep, or swine.

But Smith's bull calf will receive the first premium, and a good deal of praise gratis. There's no doubt in my mind about that, yet the question comes up, Will it *pay?* "Ay, there's the rub." I asked Smith,—

"How do your pigs grow?"

"They don't seem to grow very fast. Takes an awful sight of milk for pigs."

"Butter is pretty well up, isn't it?"

"Don't know; haven't sold any this summer. Been feeding the calf, you know."

That shows where the shoe pinches. But Smith has got a fine, yes, an extra bull calf, and will get the premium, and perhaps a "diplomzy." That's a foregone conclusion. The wise and far-seeing committee on "bulls, cows, and heifers," will view Smith's specimen of adipose matter with a look of wisdom and knowledge which only "committee men" can assume, and pronounce favorably, for fat goes a great ways in their opinion, and, like wealth, covers a multitude of sins and deformities.

Thus fat and size are made the criterion; and this false idea prevails extensively, as witness the decisions of too many of the "judges" at our annual cattle shows. Do not understand me as decrying the habit of carrying your best to your county fair, and of striving to raise fine stock and good crops. Far from it. It is just what I would have every farmer do; and here let me advise, even urge, every farmer to attend his county fair and cattle show; and not only go yourself, but take the family, particularly the boys, making the occasion an annual holiday. And further, do not go empty-handed, or to look at what somebody else contributes. Furnish your own mite towards filling up, and making the exhibition complete and interesting. Take along some of the best specimens from your crops, garden, orchard, and farm-yard, for comparison with those of your neighbors. That's wherein you are the gainer; by learning how others operate, and being willing to give up your own method and opinions for anything that is better.

But what I do condemn is the practice so prevalent of pampering, petting, and forcing certain animals or crops at the expense of others, for the purpose of creating a monstrosity of size and growth; and also the action of those committees who award premiums to such specimens. My neighbor Smith is not the only one who is feeding a pet calf with nearly all the milk of his cows, while his table lacks butter, and his pigs squeal lustily in the sty.

* * * * *

"A pretty good cattle show, Smith?"

"First rate. I got the first premium on the bull calf."

"Ah! Did you?"

"Yes. And the committee said he was the biggest and fattest calf ever shown on the grounds."

"That's just as I expected."

VIII.

AGRICULTURAL COLLEGES.

CONVERSATION which I lately had with James Brown, my neighbor the Squire's oldest son, in relation to our Industrial School or Agricultural College, presents so forcibly the opinions and position of our farmer boys in regard to the subject, that I am induced to give it here. James is a fine specimen of our well-to-do farmers' sons, with a fair common-school education, obtained at the district school; which, by the way, he has taught for the past two winters. He is now in his twentieth year, possesses a good constitution, and is endowed with a fair share of intellectual capacity and good common sense. Such young men are the pride and hope of our country, and make its best and most estimable citizens.

Since the agitation and discussion of the question of agricultural education, the Squire has been much interested, and it has been his wish and intention to assist his son towards obtaining such knowledge as should enable him to become a successful farmer, and the equal of his fellows in the other professions. And James is not averse, providing the conditions are such as he considers practicable. Meanwhile the years have rolled by, and he is nearly of age, and expects soon to enter into the labor of life upon his own account. He has but a couple of years or less, at most, to devote to study, and he is only one of a large class similarly situated.

Said James to me, —

"They are going to open the Agricultural College this spring, ain't they?"

"So I understand."

"What do you think of it, anyhow?"

"I think well of the idea, but I am not well enough informed as to the plans of those having charge to decide as to particulars. It is comparatively a new project, and we must be content to wait for results."

"I've had some idea of going to this college, but I want to understand more and better how it is to be conducted, and what will be required of the students. I've heard say that the course will be four, or at least three years. If that's so, then it's no use for me to talk about it. I haven't got any such length of time to spare, neither am I able to expend the amount of money required to carry me through such a course. I would like to spend a year or so more in learning some of those things intimately connected with farming, but I have no intention of entering upon the study of a dozen 'ologies, without a prospect of ever completing one of them."

"It is, I believe, the intention to introduce only those studies which have a direct bearing upon the object in view, and that practice and theory shall go hand in hand."

"Here I've been at work upon the farm ever since I was large enough to do chores, and I ought to have a tolerably good idea in regard to the various operations; what I want to learn now is, the why and wherefore — why such a thing is done so and so, why such acts and methods of operating produce such results, and the best ways of performing them, given

4

and received through the medium of actual practice, so that I can do it again if required."

"That is going to be one of the greatest difficulties in the way of conducting such a school successfully — to find men qualified for teachers. There are enough to be had who are fully competent to teach the theory, but are they capable of going into the field, orchard, or barn, and putting their knowledge into practice in such a manner as to instruct and benefit their pupils? That's the trouble. On the other hand, there are plenty of practical farmers who are thoroughly conversant with the methods of performing the various operations required to be done upon the farm; they have everything at their finger ends, but they have not the power and faculty of imparting their knowledge to others. Thus you see is required one of those rare men, in whom knowledge and ability to impart and execute are united. Such men our professors should be, and such they must be, to insure success to the institution."

"There is another point, — the amount of manual labor to be performed. If a fellow has got to work on the farm all day, or a greater part of the time, it's no use for him to think of studying to any advantage, for I've tried that. You hold a breaking-up plough half a day, or work in the hay field, or hoeing, or any other kind of farm work, and I tell you a fellow feels more like going to bed than getting a lesson."

"That's a fact; and though I am an advocate of manual labor in connection with such a college, as one of the very foundation stones, yet I would have but a few hours each day devoted to it, and then it takes more the form of recreative exercise than labor. Some institutions of this character devote two hours each day to manual labor, which is certainly little enough."

"O, yes; two hours would only be enough for exercise. Do the teachers or professors work with the scholars?"

"In some cases they do, in others the pupils work under the superintendence of the one who has the general charge of the farm. The former method is far preferable, as it gives more interest and dignity to the act."

"How do they divide the work among the pupils?"

"The best system I have heard of is where the pupils are divided into classes, or gangs, each one under the charge of a teacher or professor. One class studies and works in the garden for a certain length of time, and then they are changed to the fields, while another takes their place, and so on, all taking turns in performing the necessary stable work."

"That looks quite sensible. Let every one do his share of the labor and the dirty work; cleaning the stable, hauling the manure, ploughing, &c. But if there's a class of students who are not obliged to labor, but pay more money instead, it would spoil the whole thing. I wouldn't stay in an establishment where such privileges were allowed. It wasn't intended for kid-gloved dandies, but for workers. How much do you suppose it will cost for a year's board at the college?"

"That I cannot tell; but I see no reason why it might not be nearly self-supporting; that is, each pupil would perform sufficient labor to raise his own food. Of course they would not expect the luxuries and delicacies, but plain, substantial food. At any rate, with such necessary articles for the students as every farm-house could furnish, I think the expense would be comparatively small."

"Well, I believe I shall try it, if they have a short or partial course. There's a great many points I want to be posted on, and which I can't get out of books. They only tell how things are done, but I want to see them performed, and have it explained why they are done so. I want to understand the practice as well as the theory of farming."

IX.

APPLE TREES AND INSECTS.

Y neighbor Smith is just setting out a young orchard, and says he is "going into raising apples." In fact, Smith is a little excited on the apple subject just now; and, like a good many other worthy men, when he gets deeply interested in any subject, is apt to run into extremes. We confess that with apples at two dollars a bushel, there is some temptation to try and grow them; for, as Smith says, —

"Don't you see how they pay? and I might as well make the money as anybody."

All very true; and far be it from my intention to discourage the planting of apple orchards. In fact, we consider it a duty we owe to the coming generations, that they may pluck the fruits which are the results of our labors, as we do of those who have gone before. But our planters should not be too sanguine, nor expect to reap the rewards too soon, for they will surely be mistaken. An orchard is not the growth of one year nor of five; and though a man may reasonably expect to eat of the fruit of trees of his own planting, yet he must not expect to have full-grown trees from the seed, nor that they are to be grown without labor or care; neither will the fruit always bring sc high a price in the market.

We looked over the wall, and chatted a few minutes with Smith, while he was digging holes in the tough sward, and setting out his nursery-grown trees.

"What do you think of these trees?—two years old, over three feet high, and as straight as a candle. Ain't they handsome?"

Here he "chucked" another into a hole, and filled in deep with sods and wet dirt, finishing off with a heavy stamping down.

"There, I call them pretty fair. The nurseryman said they'd bear in six years after they were grafted, and I'm going to graft 'em next spring."

"Better add ten to the six."

"Why? Don't you think they are good trees?"

"O, these are not bad of the kind; but there are better ones down in your sheep pasture."

"Do you mean them little, gnarly, scrubby things?"

"Yes. Some of them are perhaps a dozen years old, and not over three feet high, but they are stocky."

"Why do you think they are better than these?"

"Because they are well rooted, and can be taken up with so many fibrous roots that the transplanting will not affect them much; and they are also very hardy, and changing from a poor to a good soil will cause them to thrive and grow rapidly. Such stocks are preferable to forced and pampered trees from the nursery."

"What do you think of this field for an orchard?"

"This easterly slope is good; but a part of the field needs draining. Trees will not do well with their feet in the water. And if you ever expect to raise an orchard, you have got to use a little more care in setting them out, besides having the ground in better condition than this is."

"What had I better do to it?"

"Cultivate it a few years with a hoed crop, and manure it well, only be sure you put on a good deal more than you take off in the form of crops. Then you can seed it down to grass, only keep the trees well and widely mulched."

"I've got a lot of old poor hay and straw in the barn; wouldn't that be good for mulching?"

"Yes; any light vegetable matter will answer. If you do not plough the ground, you can use your old hay around your young trees to good advantage. Such a mulch seems to keep the soil about the roots loose, open, and moist during the hot summer days."

"I saw you at work in your orchard this forenoon. Trimming it up?"

"No; June is the best time to prune. The wounds heal up quickly then. I was digging out borers, and cutting off caterpillar's eggs."

"Well, I never knew much about them, and how they operate, though the caterpillars come pretty near eating my trees up last summer, before I could get rid of them. They and the borers come from bugs, don't they?"

"Not exactly. I will give you their history

very briefly. The borer is a grub, which lives in the bark and wood of the apple tree, near the roots; and, if not destroyed, will girdle the tree, and kill it. This grub, which hatches from a minute egg laid by the parent beetle, lives in the wood three years, eating and growing every summer. In the spring of the fourth year it changes to a beetle, or perfect insect, which comes out of the tree, flies abroad in the night, and deposits its eggs for another generation of borers. Thus its life is divided into four states, or stages: the *egg*, the grub or *larva*, the *pupa*, which is the dormant state, and the *imago*, which is the perfect, or winged form, as the beetle. They are a great pest in some portions of the country."

"Can't they be hindered?"

"There is no certain preventive yet known. The only sure way is to examine the trunk of every tree, and cut them out, and kill them. This should be done twice a year, spring and fall."

"How about the caterpillars?"

"The apple-tree caterpillar belongs to a different order of insects, and when in its perfect or winged state, is a moth or miller, instead of a beetle. This moth lays the belt or bunch of eggs on the twigs of the trees, which I have been cutting off to-day. In the spring, when warm weather comes, a minute caterpillar hatches out of these eggs, which immediately begins to eat the tender apple leaves, and grows, and eats, and eats, throwing off and changing its skin several times during its life, till at last it is full grown about the last of June, when they leave the trees and cover themselves with an oval-shaped silken case, or cocoon, fastened in some sheltered place, and inside of this cocoon they change to the moth, which comes out in about three weeks, flies in the night, pairs, and the female lays her eggs for the next year's crop of caterpillars."

"Do they go through the same forms as the borers?"

"Just the same. The egg, larva, pupa, and perfect insect."

"What's the best way to kill caterpillars?"

"There are various methods recommended, which may be followed with a degree of success. It is always safe and sure to cut off the eggs in the spring or fall, and burn them. I have found a good way to destroy them, after they are hatched and in their webs on the trees, is to make some very strong soapsuds, and give them, nests and all, a thorough soaking, by means of a brush or swab fastened on the end of a pole. A soaking of strong suds is sure death to them. This method is practicable, for the materials are always at hand for use."

X.

MIDDLE-MEN.

HERE is a large class of persons in this country called *middle-men*, who operate in all communities, and who are of no benefit whatever. They are non-producers, deriving their sustenance from the labor of others, like a parasitic plant. They buy from the producer, and sell to the consumer, thus directly injuring both parties, — the one paying too high for his necessities, the other being obliged to sell his products under price. These drones in the social hive are numerous, and are particularly inimical to the farmers' interests. From various causes we are in a condition to be more easily preyed upon by these cormorants than any other profession. If the producers and consumers could be brought more directly together, so as to save the large profits of the middle-men, it would be greatly for the interests of both, and there would be less reason

for bewailing the small profits of the one, and the high prices of products by the other.

These middle-men comprise several minor classes, some stationary and others itinerant. These latter are found perambulating the producing regions wherever their keen visions scent the possibility of a "trade." One of these gentry, whose speciality is cattle buying, called upon me yesterday, while I was in the field, ploughing.

"Mornin', Captin."

"Good morning, sir."

"Heerd ye had some cattle to sell."

"I have a yoke of beeves. Are you buying stock?"

"Wall, a leetle; kinder off an' on like. Where's your oxen, Captin?"

"Over yonder in the pasture. You can go down and look at them. I am very busy today. Drive on, John."

He goes off to the pasture, and returns in about fifteen minutes. I ask him, —

"What do you think of them? fat, ain't they?"

"Middlin', Captin; only middlin'."

"How much do you make their girth?"

"'Bout seven foot; rawny-boned too."

"Why, I make them girth seven feet two, strong."

"Kinder flabby; they'll dress away a good deal."

"They'd ought to be pretty solid. I fed them corn meal all last winter and this spring."

"What's yer price, Captin?"

"Two hundred dollars."

"Two hundred! Can't go that, no how."

"Very well. Start up the team, John."

Buyer looks rather disappointed, but follows me round till we reach the next headland, when he breaks out, —

"Say, Captin, ain't that rather steep? Beef's a-goin' down. My brother come from Brighton t'other day, and says you can't hardly give it away there."

"When did he leave Brighton?"

"One day last week."

"Well, I've yesterday's cattle market report, and beef is tending upward. What makes you so anxious to buy if it is falling?"

"Fact is, Captin, I like the looks of them

oxen. I'll split the difference betwixt that and a hundred and seventy-five."

"My price is two hundred."

"Let's go down and look at 'em agin."

"Can't spend the time."

"A feller couldn't make his salt to pay two hundred for 'em."

"You're not obliged to take them."

"That off one ain't in near so good order as the nigh one. Think he is?"

"How did you know which was the off one?"

"Say, I'll go a five better."

"You have heard my price."

We have reached the end of the bout. Buyer makes a motion as if to leave. I make no remark, but set the plough in on the next furrow, when he suddenly wheels with—

"Here's your money. Dang it! I'll take the oxen."

"All right. John, drive them up into the yard. Let the team 'blow' a few minutes."

"Hain't got an old yoke, or halter, or somethin' that you'll throw in, to fasten 'em with, have ye?"

"Give him a halter, John."

"Good day, Captin. Looks as if we might have a spell o' weather soon."

As he goes down the road, he meets Smith, to whom he says, "The Captin is a 'hard cud' to trade with: couldn't beat him down a cent."

XI.

TAKING THE PAPERS.

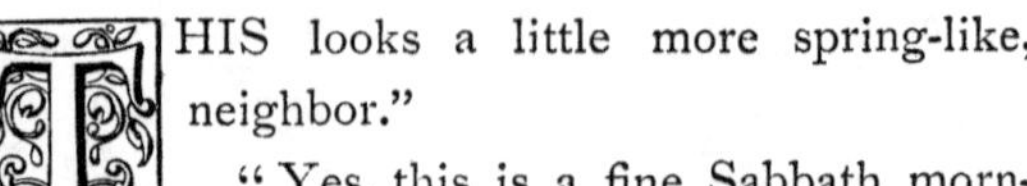

THIS looks a little more spring-like, neighbor."

"Yes, this is a fine Sabbath morning. The snow is all gone, except an occasional sheltered bank, the frost is nearly out of the ground, and the patches of new green grass we see once in a while look quite refreshing."

"Done any planting yet?"

"No; but I've got the ground ready for some early peas and potatoes."

"Don't b'leeve in putting in seed when the sile is so cold. What paper was that you're readin'."

"*The Farmer.*"

"What do you think about takin' so many papers, anyhow? My boys have been a-coaxin' me to take some kind of a farmin' paper, and I thought I'd ask you which was the best one."

"That's rather a difficult question, neighbor; but I'll give you my opinion in regard to the matter. To be without a weekly newspaper is to be far behind the times; and for a farmer to be without one pertaining to his profession, is equivalent to being behind the age in everything relating to that profession. The number of periodicals which a farmer should take, of course, depends somewhat upon his means; but no one is so poor as to be able to do without entirely. In the first place, he should support his local paper, whatever it is, if it does not deal too much in slang phrases and personalities, which, I am sorry to say, is the fact, in too many cases. Next, he should have his state agricultural paper, if there is one; if not, the one published nearest him, as that treats especially of matters around him, and is applicable to his locality. Then, a good monthly, devoted exclusively to agriculture, with a magazine for the women folks, will answer very well for a farmer's family, with such reading as is usually obtained from other sources."

"But that would be a pretty big bill to pay."

"Reckon it up, and you will find it will

come within ten dollars, including postage and all; and that is only the interest on a hundred dollars for a year at ten per cent. In what possible way could money be put to better use, or spent to better purpose, both for yourself and family? The value of a good newspaper in a family of children cannot be estimated in dollars and cents."

"There's Brown takes two or three farming papers, but he says he don't believe half there is in 'em."

"Of course, it's just like everything else; we must not believe everything we read, even if it is in the paper. There are a good many men who write for the press, who are like brother Wood, at the prayer-meeting, — they are blessed with the 'gift o' gab,' use a great many loud-sounding words, and take up a good deal of time 'a-sayin' nothin'.' It's just as hard keeping such men out of print as to keep them out of the church; but they soon die out naturally, if left to themselves."

"That's so! It isn't them that makes the most talk that knows the most — not by a long shot."

"But about reading for farmers. There are other sources from which it can be obtained without costing much. Let half a dozen farmers in a neighborhood each buy a good book, and lend or exchange among themselves. They will thus have a winter's reading very cheaply. Then there are the monthly and annual reports from the United States Department of Agriculture, which can be obtained free, upon application; also reports of State Societies, and Boards of Agriculture, when they are in operation, which can be got by applying to your representative."

XII.

THE 'OLOGIES.

WAS into Smith's last night, and found him in quite a state of excitement, consequent upon reading an article in an agricultural magazine I had lent him. In fact, Smith had "got his dander up" in regard to the doctrines and language contained in said article, and was willing to own it. He was "sputtering away" about humbugs, nonsense, 'ologies, &c.

"What's the matter with the 'ologies, Smith?"

"I don't believe in 'em, and never did. What's the use of so much 'flummy-diddle'? Plain common sense is enough for any farmer's paper. I'm a practical farmer, — none o' your science about me; and your 'ologies may 'go to grass' for all me. I've no use for 'em."

"Wait a bit; don't get excited. Let's talk the thing over a little. Nothing like keeping cool to enable a fellow to understand the gist of the matter. Now, I believe you are about as much interested in the 'ologies, as you call them, as anybody."

"No, sir; I'm dead set agin 'em."

"Let's argue the point a little."

"Wall, arger away; but fust try an apple, to clear your throat."

"Thank you; I will. Fine specimens these. What variety are they?"

"Spitzenburghs: some o' my own graftin'."

"Sure they're Spitzenburghs?"

"Sartin. I've took considerable pains to study about apples, and I guess there ain't many kinds raised hereabouts but I can tell their names as quick as I put my eye on 'em."

"I've no doubt of it; and your knowledge on the subject proves that you are not only interested in, but pretty well acquainted with one of the 'ologies."

"How do you make that out?"

"Why, *Pomology* is the science that treats of fruit, and you have shown that you know a 'thing or two' about apples; so there's one of the 'ologies."

"Well, you've got me there."

"I think I have. Hand me over that wormy apple. Do you know what made the hole?"

"A worm, of course, — an apple-worm."

"Very well; but do you know the history, habits, name, &c., of this worm?"

"No; though they say the worm comes from a miller."

"Yes, a little moth or miller lays her eggs in the calyx, or blossom end of the young apple, just as it is beginning to grow, and from that egg the worm hatches which troubles the apple so badly. After this worm is grown to its full size, it changes to a chrysalis, in which form it remains through the winter, and from which the moth comes the next spring, and lays its eggs again for some more apple-worms. The scientific name of this insect is *Carpocapsa pomonella*, and its common name is the apple-worm."

"But I don't see where the 'ology comes in."

"Yes, you can; for the science that treats of insects is termed *Entomology*. So there's another 'ology, in which you, in common with all other farmers, are deeply interested."

"You're doing well; go on."

"Well, let's go into the subject a little deeper. What's the soil of your farm?"

"Hard and rocky, mostly; some sandy loom in the low ground and flats."

"Any sand or clay?"

"Yes; there's a sand bed, and quite a clay bank over in the corner. There was a 'brick kill' down there a good many years ago."

"Any large stone?"

"Some bowlders—granite; and the ledges crop out a little over in one corner of the pasture lot."

"Very well; there's another 'ology—*Geology;* which treats of the formation and structure of the earth, and of what it is composed."

"Well, I'll give up. Looks a little stormy out, don't it?"

"What makes you think so?"

"Because there's heavy rain-clouds rollin' in from the water, and the wind is 'out' strong."

"That's another 'ology—*Meteorology;* which is the science that treats of the weather and the condition of the atmosphere; a science with which every farmer is more or less practically acquainted. These are only a few of the 'ologies in which every farmer is directly interested, and the principles of which are so frequently brought into his every-day practice. It's no use, Smith, for you to be 'down' on the 'ologies."

"Can you name another?"

"Certainly. There's *Physiology*, both animal and vegetable; the former treating of everything that relates to animals, and the latter of plants. So you see it is absolutely necessary for a farmer to know something about this 'ology. The sciences are intimately connected with the farmer's operations at every step of his progress, and he is not a wise man who persistently opposes whatever to him seems to smack of science, or is comprehended under the heading of an 'ology."

XIII.

AN EVENING'S CHAT.

"THIS looks a little winterish?"

"Yes, decidedly so. Pretty good sleighing now, isn't it?"

"Pretty fair; a day's travel will make it excellent. Isn't this earlier than usual?"

"Rather. There isn't more than one year out of ten we have sledding by the first of December."

"We haven't had any very cold weather yet?"

"O, no; the coldest was this morning, when the mercury was down to 15°; cold enough, though, to give us warning to prepare for it."

"I was thinking of going to market to-morrow with a load of potatoes. Do you suppose they will freeze?"

"Not unless it is colder than it is now. I hauled potatoes, in barrels, last winter, at any time when the thermometer indicated a temperature above 15°, without any danger of their being frost-bitten; but for a long distance it wouldn't do to run the risk. It is said it takes several degrees more of cold to freeze them when in motion than when at rest."

"Potatoes are doing pretty well in market?"

"Yes. Footes and Jacksons bring fifty-five cents; and at that price we can make a fair profit raising them, providing they don't rot; but if they do, it kills the profits."

"It's no use to plant any but these sorts for sale or family use. There's the Californias turn out surprisingly, but you can't sell 'em in market. They're nothing but a 'bag of water' cooked—only fit for stock; and I am doubtful whether they pay then. They are handy, though, to have to feed occasionally through the winter, to keep the cattle loose. I'm feeding mine out now with my straw and coarse fodder, which I like to get rid of before mid-winter."

"Well, Californias are poor, ain't fit to eat, as you say; but I cooked and fed out twenty-five bushels this fall to my pigs, and they did well on it."

"Guess you give 'em somethin' besides the potatoes."

"Of course. Mixed them with corn meal."

"I see you've butchered. How much did your pigs weigh?"

"One went two hundred and thirty-two, and the other three hundred and two pounds, at seven and a half months old. That sow was the largest and fattest pig of her age that ever I raised. They got nothing through the summer but the waste milk of three cows and the slops from the house, and were fatted on potatoes, oatmeal, and bran, and corn meal for the last month. They both received the same keeping, yet the sow gained the fastest. I have always found that sow pigs are best for fall killing, but barrows are better to keep over winter."

"I s'pose you'll keep a spare-rib for Christmas?"

"Yes; I have packed them into a nice, clean barrel, with snow. They have frozen, and will keep first rate; but I guess some of us will have to do without mince pies, apples are so scarce, or make them minus the fruit."

"I've got one barrel of Baldwins headed up and put in the cellar, and I shan't open it till

Christmas. If we have many more such cold, hard winters, I shall lose all my apple trees."

"Don't lay it all to the cold; it isn't exactly that which kills the trees, but it's the sudden changes, or extremes of temperature, usually in March. Thus we have a day or two of quite warm weather, which swells the buds and starts the sap up the trunk of the tree, then the wind suddenly changes to the nor'ard, the mercury in the thermometer goes down with a jerk, the sap freezes, and the buds are killed or the trunk bursts open. However, if we experience no sudden changes this winter, and no frost in blossoming time, we shall undoubtedly have a heavy crop of apples next year, for the trees had a good season last summer to rest and obtain food for forming fruit buds for the next season's crop."

"I'm going to graft some more of my trees next spring. What kind should you put in?"

"Mostly Baldwins. There is no variety that has yet proved equal to them for a general market apple in this latitude. But there is one thing I would advise you not to do, and that is, not to graft your old trees if they bear a fair marketable apple, because it won't pay. A good many pretty good trees, and not a few orchards, have been ruined by their owners getting a touch of the 'grafting' mania."

"That's a fact. I spoiled three or four. I'll never graft old, large trees again. But I must get home and put up my stock. Good day."

"Drop in again, neighbor, some of these long evenings, and we'll have a social chat about matters and things in general, and farming in particular."

XIV.

PLANTING FOR POSTERITY.

"SETTING out shade trees?"

"Yes. Planting for posterity."

"Hem! And posterity won't thank ye for it. Let every one look out for themselves, I say. I shan't trouble myself about what will happen fifty years after I'm gone."

"To tell the truth, neighbor, it isn't wholly for posterity after all; there are some selfish views at the bottom."

"You don't never calc'late they'll grow big enough to make any shade worth mentionin' in your day, do ye?"

"Perhaps not; but then they may. Do you see those Balm o' Gileads that are higher than the barn now? Well, twenty years ago they were nothing but sprouts, which I stuck into the ground."

"Yes; but they are fast growers, and don't amount to anything after all."

"I know they are not a very valuable variety for lumber, but it is a rapid grower, and easy of propagation, and while young makes a fine shade tree."

"How old's them horse chestnuts?"

"Ten years from the seed; and now they are about twenty feet high, and over six inches through at the butt. Last year they were covered with cones of splendid blossoms, making a very pretty sight, I can tell you."

"But they're in a rich place."

"Every tree transplanted should be placed in a good soil, adapted to its nature, else it's no use to set them out."

"It costs something to take so much pains."

"Yes, it costs, and it pays too. It pays not only in the pleasure of their possession, but in real dollars and cents. How much more do you suppose your farm would be worth if you had a row of handsome, well-grown shade trees on each side of the road through it?"

"I don't know as t'would be *worth* any more."

"Well, then, how much more would it *bring*, supposing it was put into the market to-day?"

"Something, I s'pose."

"Yes; a pretty large 'something.' It would bring enough more to pay for the time, labor, and cost of setting out and taking care of the trees, ten times over. There is no method of adorning, and at the same time adding to the value of a farm so cheaply, as by planting fruit and shade trees."

"Well, perhaps it's so."

"Then again, as to planting trees for posterity. I consider it my duty as well as my pleasure to do so. Not only so because those I have set out in years a-gone have repaid me so well, but as a debt I owe those who, many years since, planted these trees which now give me shade and fruit. I will endeavor to do for those who come after me what my fathers have done for me. See those noble elms, some of them of nearly a century's growth! Supposing our forefathers had decided that it didn't pay to plant trees, where should we have found such specimens? What better and more enduring mementos and proofs of their love for their race could they have chosen?"

"I suppose the elm is the best tree for shade there is, isn't it?"

"It is certainly the very king of trees; but there are many others worthy of a place beside it. There is the rock or sugar maple, which cannot be surpassed as a hardy, thrifty, and fast grower, with a fine form and foliage. The white or soft maple is also a fine tree, and especially in the autumn, when its brilliant colored leaves make it a very beautiful and desirable ornamental and shade tree. Then there are the ash and the oak; good shade trees, but rather hard to transplant. Also the beech and the birches, which are very graceful, though but little used."

"What's the best time in the year to transplant trees, do you think?"

"In the spring, as soon as the ground is dry enough. At any rate before the buds begin to burst."

"Do you believe in cuttin' off much of the top when they are set out?"

"That depends upon circumstances. If the roots are all, or nearly all, obtained, but little top pruning will be needed; and, if possible, only those trees should be taken up which can be removed with most of their roots. However, if it is thought advisable to transplant a larger tree, and many of the roots are lost, a severe top pruning is necessary."

"Had the soil ought to be manured before setting out shade trees?"

"It is not necessary if it is in good condition. But all newly-transplanted trees should be deeply mulched the first year at least, to keep the soil moist, and the roots from suffering for moisture. There is another point in which many tree-planters err. They dig a hole deep in the ground, thrust in the tree, cover its roots deep with earth, and think they have done their part. Now, unless the soil is very dry and gravelly, the tree had better have been set on the top of the ground, or nearly so, and the roots covered with soil taken from another place. Trees should also be firmly staked."

"How about firs, spruces, &c.?"

"Evergreens require a somewhat different treatment from the hard-wood trees. They are harder to live, and require more care in transplanting. It should be done in early summer, or when the buds are swelling. Some are very successful in taking them up in the cold season, when a large lump of frozen earth can be taken up with the roots. They should never be pruned, or at least their lower limbs should not be cut off, as it is their nature to grow low limbed when standing alone."

XV.

PARASITIC PLANTS.

OOD morning."

"How d'ye do, Squire? Walk in; take a chair, and draw up to the fire."

"Pretty cold, ain't it?"

"Yes; ten degrees below zero is about cold enough. Do you suppose this cold snap will last long?"

"No, I guess not; the wind is haulin' so'westerly, and it's 'bout time for our January thaw."

"Help yourself to an apple, Squire."

"Sweet, ain't they? What do you call them?"

"Talman Sweets. They are the best winter sweets I ever raised. Splendid for baking, and first rate to eat any way. I've just been examining this one with the scabby-looking spot on the side."

"What do you make it out to be?"

"It's a plant that grows on the apple."

"A plant! Why, how does it grow there?"

"Like other plants; only it derives its nourishment from the apple instead of the soil. Such plants belong to the lowest orders of the vegetable world, and are called *parasites*, which means a kind of plants which grow on and from other plants, to which they attach themselves, and whose juices they absorb."

"How do they get there in the first place?"

"These parasitic plants are propagated by means of organs called *spores*, which correspond to the seeds of other plants. When this parasite obtains its growth, it is covered with these spores, which are so minute as not to be readily discernible by the unassisted eye, but which are plainly shown under the microscope. They are scattered abroad by the wind and other agencies, and when one of them effects a lodgment on some plant which is adapted to its growth, it immediately takes root and flourishes if surrounding circumstances are favorable."

"I should think it would injure the plant or fruit on which it grows, as you say it sucks out the sap or juice for food to live on."

"Certainly it does. Look at this apple.

Don't you see there is a depression, or hollow on the side where the parasite is situated. This is caused by the parasitical plant drinking up and appropriating the juice or nourishment which should have formed the flesh of the apple. Some of these parasites appear only as minute spots, no larger than a pin-head; others are as large as a dime, and others still covering a quarter, or perhaps a whole side of an apple, producing a gnarled and withered appearance."

"Are these parasites found on all kinds of plants?"

"Probably every species of plant has its parasite, which is, or may be present when the plant is in a condition to sustain it. I do not suppose a perfectly healthy plant would be thus affected, but the presence of a parasitic growth is evidence of disease and degeneracy."

"Are large plants, like trees, troubled the same way?"

"Yes. For instance, the common cherry and plum tree is affected with a parasitic disease, popularly known as the *black-knot.*"

"Why is that a parasitic plant? I always thought worms caused the black-knot."

"No. It is a parasite, which scientific men call *sphæria morbosa*. It is propagated by spores, as I described to you before, and has proved very destructive to the plum and cherry throughout the country."

"What time of the year do you calc'late the seeds spread that make the plant?"

"In late summer or autumn, or when the plant has arrived at maturity. During autumn, or perhaps winter, the spores of the parasite are free, perhaps floating in the air; they lodge in, or rather on, a branch or twig of a tree, which is in a suitable condition for retaining and nourishing the plant. There they remain, embedded in some minute crevice in the bark, until spring, when the warmth germinates the spore, which, swelling, causes a slight irritation in the bark, and the plant roots and grows. As it grows, it appropriates a portion of the sap which is circulating in the twig or limb to its own use, and spreading out its roots, changes the growth to a mass of spongy texture, swelling, bursting the bark, and continuing to enlarge, until all the sap in the limb is required to support the parasite, when the extremity of the twig withers and dies. It is now, perhaps, midsummer, and the parasite, or as some call it, wart, excrescence, or black-knot, is of a greenish color on the outside, and it has become the habitation of numerous 'worms' or larvæ of several species of insects, which have found its soft, spongy body a fine place for breeding purposes. Hence arose the idea that the knot was caused by insects. The plant has now nearly got its growth, and the outside of the knot begins to turn brown, and by and by, at early autumn, is black, and covered with the spores of the perfectly developed plant. It is fully ripe, and ready for entering upon the great life labor of all animate things — the perpetuation of its kind. It is now coal-black, and the spores may be seen under a microscope. They are very minute, ripe, and ready for distribution."

"Ain't there any cure for the black-knot?"

"None as yet known and practised. It could be prevented, but it is hardly probable it ever will."

"How could it be prevented?"

"By destroying the plants or parasites before they ripen or go to seed and fully develop their spores. If the affected twigs and limbs were cut off and burned in early summer, while the parasite was yet growing, they would be destroyed for good, and so much done towards 'heading it off,' or preventing its further spread. But such a course, to be effectual and practicable, would require to be thoroughly performed throughout a whole section of a country. It is useless for one person to try to save his trees while his neighbor's garden is full of trees burdened with the unsightly knots from which the spores may be scattered by every passing breeze."

XVI.

ROAD-MAKING AND BREAKING.

"WHO d'ye s'pose they've put in road surveyor this year?"

"Can't guess. Hope it's somebody who knows enough to turn a straight furrow. Who is it?"

"Same as last year."

"Is that so"?

"Yes. I guess they're going to make a life member of him."

"Well, if they are, I hope they'll keep him on the superannuated list."

"I don't see what the selectmen are thinking about. Here's our roads been goin' from bad to worse for the last three years. It's time we had a surveyor that knew something about making a road."

"Perhaps he'll do better this year."

"Better! Why, he thinks he's the greatest road-builder 'out.' Look at that piece of road on the flat down below my house. He went and ploughed the chimes off last year, and scraped the dirt into the middle, leaving the road so narrow it's almost impossible for two loaded teams to meet and pass each other without running into the ditch or tipping over. If it had been a rocky, hard place, there might have been some excuse; but there was plenty of room to plough, and enough dirt outside of the old ditch, — and conscience knows the road was narrow enough before, — but he must plough off the chimes, and make it narrower still."

"I've noticed that piece of road."

"Then there's the top of the 'brook' hill; — he went and throwed it up last fall, making the hill at least a foot and a half higher, — as if it wasn't hard enough to haul up before, — instead of ploughing and scraping it off, and thus making the hill a little lower and easier every year."

"That's a common fault with our road-builders and repairers. There are many small hills and elevations, or 'rises' in the roads, which, if ploughed and scraped down into the hollows — a little being done each year — would soon be on a level, and the road become gradually graded. But ignorant surveyors plough and throw up on these elevations just the same as they do on the lower parts of the road; and, worst of all, they cannot be made to believe but their 'way' is best."

"A main travelled road should always be wide enough for loaded teams to meet and pass each other without any danger of running the off wheel into the gutter or of tipping over. Of course it isn't expected cross roads, and those but little travelled, should be made so wide."

"There is another thing in which our surveyors are very negligent, which is, in seeing that the main roads are properly broken out in the winter, after heavy snow storms."

"That's a fact. See how our surveyor broke out our road last winter! He put on a single ox-sled, thus making the road just wide enough for one team and no more. On the plains, where it was drifted badly for half a mile, he shovelled a regular canal the whole length, just as wide as one ox-sled, and only two or three places for teams to meet and turn out in the whole distance. It's disagreeable, especially in a cold day, to have to wait five or ten minutes for a team to come up before you can drive on; and worse than that, every time there came a light snow storm it would drift in full, or nearly so, and have to be shovelled out again."

"That's no way to break roads. The great point is always to keep on top of the snow, and make the track wide. It should be trod and broken down under foot instead of shovelled out, in all cases where a team can be got through. Then it will not cost so much to keep the roads in order during the winter season, for they will not drift full every storm, but most of the snow will blow over."

"Nothing like a good, large, old-fashioned triangle, with a big team of oxen hitched on to it, to break out roads with. Haul it twice over a road, no matter how heavy the storm was, and it will crush down the snow solid, and leave a track wide enough to let teams meet and pass almost anywhere, without getting the horses or oxen floundered in the snow."

"Yes, that's the best method I have ever seen employed. It's one of those old-fashioned ways of doing things which is preferable to modern methods."

"There's only one thing about it. If our roads ain't kept in better condition for travelling this winter than they were last, I shall complain on 'em."

XVII.

IN THE BARN.

EED your cattle pretty often, don't ye?"

"Five times a day."

"Hay every time?"

"No: hay in the morning as soon as I am up, then give part of them roots or grain after breakfast, then water, and leave most of them out in the yard if the weather is fair. Feed them all at noon with coarse fodder, if I have it; if not, give them poor hay. Water, stable, and feed them again at sunset, and a light foddering before I go to bed."

"Feed your sheep and horses the same?"

"No. Three times a day is sufficient for them, unless you are working the horses hard, then they should be fed more."

"Water 'em all twice a day?"

"Yes. I consider plenty of good water essential to the health and well-being of all kinds of farm stock, and no farm-yard should be without it."

"Well, I know it's handy to have a plenty of water for stock, but some farmers never give their sheep any, and say they do first rate."

"I know some pretend to say sheep need nothing to drink in the winter, but I don't believe any such doctrine. I know sheep will not drink so much water, accordingly, as other animals, but I think it's best to have it where they can get it handy, at least once a day, if they are so inclined."

"Mine eat a good deal of snow."

"Got any lambs yet?"

"Jest beginnin' to come along. I find it's 'bout as much as I want to do to take care of 'em this cold weather."

"Now is the time sheep and lambs require special care."

"What do you do with 'em when they're chilled?"

"Take them near a warm fire, and wrap them dry and warm. A good many lambs are lost for lack of a little attention at the right time."

"Ever feed any cow's milk to 'em when the dam don't give enough, or won't own 'em?"

"Always; and save many that way. It should be given warm, by means of a false teat. Use new milch cow's milk, if possible; but if you have only farrow cow's milk, it will answer, if sweetened. I remember once, years ago, of bringing up a pair of twins, which the mother wouldn't own, wholly on cow's milk, by suckling them regularly on the cow. They grew and throve well, but it was costly, and a good deal of trouble. If a lamb lives to be four days old, and sucks well, there isn't much danger but they will do well."

"What kind of grain is best for sheep?"

"I like oats best, though beans are first rate."

"What do you think about raising twins, anyhow?"

"I think it don't pay, unless the dam is large, and has an extra flow of milk. One lamb is enough for a sheep."

"Got your fodder all fed out?"

"Nearly. There's a hundred or two more of fodder corn, which I keep for baitings. Did you raise any of it last year?"

"No. Cattle like it?"

"First rate. Mine eat it up, leaf and stalk, and look for more. With a little pains, a farmer could raise a big pile on an acre."

"How did you plant yours?"

"In drills, pretty thick."

"Southern corn, I s'pose?"

"Yes; or rather western—the large 'horse-tooth' variety. I turned over a piece of green-sward after I had got done planting, harrowed, drilled, putting a little manure in the drill, dropped and covered the corn, giving it a light top-dressing with plaster as it broke through the ground, cultivated it once, and raised a fine crop."

"How did you save yours? Some find it hard to keep it from hurting."

"Yes, I know it; and it requires some care, the stalks are so succulent and full of juice. I cut mine up, let it wilt thoroughly, tied it in small bundles, hauled it into the barn, letting it lie on the rack over one night to sweat a little, then hung it on poles in the top of the barn, and it cured nicely."

"Feed out any potatoes this winter?"

"A few; just enough to keep the stock 'loose,' and to counteract the dry hay. A mess of potatoes twice a week is enough. When potatoes will bring fifty cents a bushel in market, and corn meal can be bought for a dollar, or thereabouts, I'd rather feed the corn; — meal for the cattle, and corn for the horses, and a little to the sheep."

"Use bows to tie your cattle with?"

"Part of them, and some chains."

"What do you think of slip stanchions?"

"Don't like 'em. They are too rigid; don't give the cattle chance to move and rest or stand easy. After all, there's nothing better or surer than the old-fashioned bows. A set of good, strong, easy-working bows and wreathings is safe, easy for the animal, and permanent. Ropes are unhandy to tie, chains will unfasten sometimes, and slip stanchions are hard for the cattle."

"How do you like your new shed over the manure pile? S'pose 'twill pay?"

"I like it well. It helps keep the stable warmer, and the manure does not fill full of snow in winter, nor leak away with the spring rains. I think, too, it's a good deal stronger when put on the land. Besides, I have more of it; it don't blow nor wash away, and the cart-loads of muck which I put at the bottom of the pile in the fall come out the next spring as good manure as the rest of the pile."

"My calves are getting lousy. What would you do to kill 'em without hurting the calves?"

"Take some clean soil, dry it thoroughly, pulverize fine, and rub it into their hair; or rub in some grease, or lard, or whale oil; most any kind of oil is death to lice. Kerosene is apt to hurt the calves, so is tobacco; and lime or ashes, as some use, is sure to take the hair off."

"Is them two-year olds of yours broke?"

"Yes; the boys always handle all the steers when they are yearlings, and by the time they are three-year olds they are thoroughly broken, and drive like old oxen. That's the way to do it; not wait till they get their strength and growth, when it takes a strong man to do anything with them, and they are more likely to be cross too."

"Cows come in yet?"

"Only one; expect another to calve soon."

"Goin' to raise 'em?"

"Certainly. Always raise the early calves. They are worth more to keep for any purpose."

"Give 'em all the milk of the cow?"

"No. Take them from the cow after they clean the bag thoroughly; give them some new warm milk for a few days, and gradually change them on to skim milk and meal, fed in a trough. Once learn them to drink, and they will do well, if you give them enough to eat. An early calf will begin to eat hay by the time it is four weeks old."

XVIII.

HOW TREES GROW.

SEE, neighbor, you have cut down the old apple tree that stood before your door so many years."

"Yes. I told the boys this morning they might cut it down for fire-wood. It was long since past bearing fruit, there being only one live limb on it, and I've been threatening to cut it down the last three years; but somehow I didn't like to destroy it, for memory's sake."

"How old *was* that tree?"

"I don't know for a certainty; but it was nigh on to fifty years old. I made it out to be forty, but it was so rotten at the heart I couldn't exactly tell."

"Some say they can tell just how old a tree is. They cut a mast over in Tunk the other day, that they said was a hundred years old."

"O, yes, it's easy telling how old a tree is,

because it makes a layer of wood on the outside next the bark every year that it grows, and so by counting the layers from the heart to the bark, you can find out its age. The sap or food that the tree lives on comes from the ground, passes up through the roots and trunk or wood of the tree to the leaves, where, by coming in contact with the air, it undergoes certain changes, which makes it fit to form wood of; it then goes back down the tree, principally between the bark and wood, and as it passes along, it slowly hardens, and forms the annual thickness of wood. You know the pine 'sliver' that the boys like so well in the spring; well, that is the sap or wood-making substance just returning down the outside of the tree, and in a few weeks it hardens, and forms the yearly ring."

"Yes, I understand; and I suppose that's what makes the top or ends of the limbs grow the fastest. When you graft a limb, does the sap act just the same?"

"Yes; the sap passes up through the scion to the bud or leaf, and then back or down, and if the sliver or bark is nicely joined to that of the stock, the sap continues its course, and the graft will live. But if the stock pinches the scion, it will crush the pores of the wood, and the graft won't 'take' and live, neither will it unless the bark joins."

"I don't believe in grafting large limbs — do you?"

"No, I do not. Here's a good example of its bad effects in the tree I cut down this morning. It was grafted about eight years ago, and some of the limbs were cut off where they were four inches through, none of them being less than two inches in diameter. To be sure the scions lived, but the wounds never wholly healed, the bark on the upper part of the limbs peeled off, and the tree decayed gradually."

"What do you calc'late makes the bark die on one side of the limb and not on the other? I've noticed it in a good may trees in my orchard."

"It is never so only when large limbs are grafted. Cut off and graft small limbs, and no portion of the bark is ever injured. The cause of the bark dying so, is, that there is not a sufficient number of boughs on the graft to make, or rather prepare, sufficient sap or food to form a layer of wood over the whole limb; and again, as the leaves are the lungs of the tree, there was not enough of them to draw up sufficient sap through the wood for the full support of the limb, hence some portion of the bark had to wither and die for lack of nourishment; and the reason why it is the upper side, is because of the influence of the sun on that portion."

"But how does the sap get started in the first place?"

"That is a question which vegetable physiologists are not yet decided upon. One of the most reasonable theories advanced is, that the warmth of the sun's rays in the spring causes a swelling, or irritation, creating a vacuum in the bud, which results in setting in operation the powers of capillary attraction."

"I noticed, when I was cutting fire-wood in the woods the other day, when that warm spell was, that the sap had started considerably in the beeches, maples, and birches."

"So did I; and you remember the wind suddenly changed, and it grew very cold before the next morning. That's what's bad for apple trees, — having a warm spell, just enough to start the sap, and then a sudden change to severe cold. It was that which killed so many orchards in the year of 1856 and '57."

"Got your summer's wood most cut?"

"Nearly. I shall get up a cord or so more, and then there will be enough to last me till next January."

"Nothing like having plenty of fire-wood on hand."

"That's so! There's Smith, has to get up in the morning in haying time, and cut wood to get breakfast with. A farmer never can get along that way. He ought to get all such things done up during the winter and spring, when he is not driven with labor, so that in the farming season he can be all ready for work."

"Cutting your second growth clean?"

"No. I cull out the poorest, and where it is the thickest, taking a piece each year, so that part I first commenced on is now the largest and handsomest in the lot."

XIX.

PIGS AND POULTRY.

HAT you getting out timber for? goin' to building this spring?"

"I think of putting up a building for a hog-pen and hen-house, if I can get time before planting. Shall build it ten feet posts, and use the lower floor for the pigs, the upper part for a hennery. What do you think of the plan?"

"Seems to me you're taking a good deal of pains for the hens. Mine stay in the barn."

"I don't like to have hens round all over the buildings, especially in the winter time, when they waste about as much hay as they are worth. And they want a warm place and good feed, with enough of it, else it don't pay to keep 'em."

"Does it pay anyhow?"

"Yes. I have proved to my satisfaction that it pays well to keep a small lot, say twenty-five, for the eggs they produce. With eggs at a quarter of a dollar a dozen, as they have been the past year, it don't take a great while for that number of hens to bring in a ten-dollar greenback. It isn't a great deal of labor to care for them, and in the summer time they will pick up more than half of their living."

"Yes, and scratch up the garden into the bargain."

"It's easy enough keeping them shut up a week or so in planting time."

"What breed are yours? some of the fancy kinds?"

"They are no particular breed, but are such as we have kept on the farm for years. They are medium-sized fowls, yellow-legged and flesh, hardy, and good layers. It has been our custom to change the cock about once in three years, getting one from another flock, and some pullets are grown every year to replenish the number, as we never keep a hen after she is over two years old."

"Do yours lay all winter?"

"No. We generally give them a resting spell of six weeks or two months in November and December, and they do all the better for it the rest of the year."

"Somehow I never could make my hens lay of any account in the winter. I feed 'em all the grain they want, but the eggs don't come."

"Hens require something besides grain to make them lay well in the cold season, and when the ground is covered with snow. In the first place, they must have material for making the shell; grain does not furnish enough, and their eggs are soft-shelled. Give them burned clam or oyster shells, or old plaster, or anything that contains lime, and the egg-shells will be all right. Then again they should always be provided with a large box of clean, dry sand and gravel, with a few wood ashes to dust in, for killing and preventing the lice from injuring them. The gravel is also necessary for keeping their digestive apparatus in order."

"I know it's queer how much gravel a hen will eat. Their gizzards are full of it."

"Another point. In winter, hens must be fed occasionally with some kind of meat, or animal food, in addition to their grain. There is nothing better than the crumbs and bits from the table. In the warm weather they will provide themselves with this sort of food, and they thus destroy and devour large quantities of insects in the fields and gardens."

"Shall you have your hens roost in the room you are fixing up for them?"

"Yes. I shall put in some poles on one side, and also save the droppings, which makes the very best kind of manure, you know. I want this room for them particularly in cold weather; when it comes warm, let them run out. Shall put in a large double window on the south side, so they may have sun in the cold days. Light and warmth is what they want."

"Got any pigs yet?"

"One litter; expect another in a few days."

"Pigs have been pretty high for the last two years."

"Well, yes; five dollars apiece is a pretty big price to pay for a pig four weeks old."

"How many have you got?"

"Eight. One died, and the sow killed one."

"What do you do to the sows when they begin to kill their pigs?"

"Get them drunk as soon as possible, and then there isn't any more danger. They seem to forget their cannibal propensities after they get over their drunk. Give them a pint of rum or whiskey, or a quart of hard cider. If they won't drink it readily, mix it in their meal, and it will answer the same purpose. That's the only sensible use I ever saw rum or whiskey put to."

"Some sows are apt to lay on their pigs and kill 'em."

"That's sometimes the case with large sows. To prevent it, place some poles around the pen, about a foot from the floor, and the same distance from the wall. That gives the pigs a chance to keep clear of getting crushed when the sow lies down."

"Going to have wooden floors in your hog-pen?"

"Yes; plank. I believe in keeping hogs dry and clean, or at least their sleeping-places. I like to have them get to the ground, and always have a yard adjoining the pen, where they can run in and out at pleasure during the warm weather. You can also make several cart-loads of the very best kind of manure, if you keep the yard and pen supplied with muck, brakes, straw, and other vegetable rubbish."

"'Twont pay to keep many pigs with corn at a dollar and fifty cents a bushel."

"That's a fact. Neither will it pay to throw away the waste milk of three or four cows for want of pigs to change it into pork. Whether it will pay to do such and such a thing depends so much upon circumstances that it's difficult to decide in all cases."

XX.

FARM FENCES.

THIS question of fence or no fence is becoming a matter of serious discussion among the farmers of our country. Where land is divided up into small farms, as with us, the cost of building and maintaining fences amounts, in the aggregate, to an enormous sum. In fact, it has become one of the largest items in the small farmer's bill of expenditures. Now, is all this outlay necessary? We think not. We have no reason to doubt but that at least half of the farm fences in this country could be dispensed with, and a large portion of them need never have been built. This fence-building illustrates one of the natural tendencies of the age. We are far too exclusive as a people, and as a class of peoples. We fence round about, and hedge ourselves in, not only in so far as land is concerned, but also in our social relations and our lives. We fence our farms from the roads, and our houses from the streets; fence our fields across and through, into squares, triangles, and angles which do not admit of a name, and all without adding any real value or convenience. Our road-side fences, built huge and high, seem to say, "Thus far, and no farther;" and all our out-door farm fixtures and appointments seem to wear an air of privacy and exclusiveness, strongly in contrast with the farmsteads of other lands. Of course, a certain amount of farm fence is absolutely necessary, as around pastures for stock, and generally on boundary lines; but further than this they may be dispensed with, and their cost applied to some other purpose. Let it once become the custom for each farmer to fence only to confine his own stock, and not against his neighbors; and let every state enact laws against cattle running at large; then only pasture fences will be needed. The object in fencing should be to keep stock in, rather than to keep it out.

"Getting out fencing stuff?"

"Yes. I'm going to try some more of the stake and wire sort."

"What I've seen of it seems to stand well."

"Mine does that I built last spring. I think it's the best kind of fence you can find for clay soils and low lands, where the frost heaves bad. If the stakes are hove out, you can easily go over it every spring, and drive them down again."

"How did you build yours?"

"I used a crowbar to make the holes, then drove in the stakes and wired the tops. I set mine a foot apart, making sixteen stakes to the rod. Old growth cedar, split, makes the best stakes; the saplings rot down in a few years. I shave the tops off smaller, so it won't take so much wire to go round them."

"Makes a cheap fence — don't it? What did yours cost a rod?"

"I can get stakes at a cent apiece; sixteen to a rod would be sixteen cents, and the wire would be about sixteen cents more, making two shillings a rod cost for materials; — the cheapest kind of fence, at that rate, that can be built."

"Nothing like stone wall to last."

"Yes. And no doubt on rocky farms, and

where the material is handy, it is the best and cheapest in the end. A stone wall, well built, of the proper shaped stones, will stand a good many years on upland and dry soils; but it's no use to build it on clay soils, for it will tumble down, and have to be laid up over again every two years. Cedar is the material for fence on such land."

"What do you think of posts and rails?"

"Make a first rate fence, even on clay ground, if the posts are set in good cedar feet. Pitch-pole fence ain't of much account, and pickets, with iron posts, are not lasting. Board fence don't stand well, but answers pretty well, after all."

"Wall ain't fit to stop sheep."

"No. Board fence is best for a sheep pasture, and if you have a wall, it must be well top-poled. How much do you calculate it costs you annually to keep your fences in repair, and build such new ones as you think are needed?"

"Couldn't tell; pretty big sum though, I guess, come to reckon up all the time and labor."

"Yes; it's larger than we farmers have any

idea of, and it's my opinion we lay out altogether too much in fence-building. I believe we could get along just as well with half as many. They are not only a great cost, but in many cases they are an inconvenience. Take, for instance, our fields where we do most of our farming. Nine out of ten of them are cut up by fences into lots, varying from one to ten or more acres. Did you ever notice how unhandy it is to work in a two-acre field, or any small field, bounded on all sides by a fence? It causes a great loss of time at the headlands in ploughing, furrowing, or whatever is being done, which would not be so if the fence was not there. Then again, there is the loss of the land on which the fence stands, and a strip along the side, which, do the best you can, will grow up to weeds and small bushes. It's about as bad after the field is seeded to grass. It takes longer to cut the hay in an acre field fenced, than it would to cut an acre taken from a large tract unfenced."

"You've always got to have good fences round the pastures, to keep the stock in."

"Certainly; and that's the only place where a fence is absolutely necessary."

"What would you do about the roads?"

"A fence is not absolutely necessary even there, if the pastures are well fenced, and every man is obliged to take care of his own animals, or pay fully for what they destroy. I have seen here whole farms, bordering on the main road, without a rod of fence to protect the fields and crops from passers by; and growing and ripening crops planted up even with the roadside, and nothing injured or harmed. In fact this is becoming a general practice in many sections where fencing material is scarce, and I hope the system will be followed. In the old countries, where every foot of land is required for some purpose, and is valued accordingly, there are no fences; the fields are ploughed up even with the roadsides, and crops are cultivated by the sides of the walks, and nothing is harmed or destroyed, either by man or beast. It is the custom. They are protected by the laws, and the crops are safe."

"We ain't so hard on it for land in this

country, and I guess such a way of doing things wouldn't work very well here."

"I know that one of the causes that have led to so much fence-building with us has been the abundance and cheapness of land; but as the population becomes more dense, and the soil more generally occupied, land, particularly in the older settled portions, will become more valuable, and a general fence-destroying practice will be adopted; and I have no doubt but in twenty-five years, farmers in all parts of the country will dispense with, at least, one half of the present number of fences."

XXI.

OUT IN THE FIELDS.

HE Sabbath is peculiarly the farmer's day of rest and recuperation from the severe labors and strain upon the physical system during the week, and particularly in the growing season, or from the time he places the first seed in the bosom of old mother earth until its yield has been put away for safe-keeping in his cellar or granary. Spring, summer, and autumn are seasons of protracted toil and continual care, which is only partially relieved by the winter's cessation from active duties, and the seventh day is to him a welcome day of rest. But few farmers, comparatively, are so situated as to be able to avail themselves of religious privileges, so far as relates to attending regular public services. His home duties require attention, which must be given upon all days. His chores must be done, his stock cared for, and his growing and ripen-

ing crops carefully watched, lest in an hour he should lose the result of many a hard day's labor. But he is not entirely wanting in religious privileges and advantages. The very nature of his occupation is ennobling, and provocative of thought and feeling which acknowledges the presence of a superior power. The various operations of nature continually going on before him — the springing of the seed, the flowering season, and the ripening and decay — all tend to suggest thoughts as to the why and the wherefore, which can only admit of one answer. If he does not read sermons in stones, he does see and acknowledge an infinity in everything. He can never be an atheist.

To me there is nothing in nature so beautiful as a clear, calm Sabbath summer morning in the country. The quietude and stillness in the very air, so different from the bustling sound of the week-day; the absence of noise and commotion about the farm-house; and then, when the church bells of the distant villages sound out softly upon the odorous air, a sense of peace, and security, and of a superior Presence, seems to fill and pervade all things. It is such morn-

ings as these that the farmer thoroughly enjoys. He has performed his various out-door duties, looked over his weekly papers, and, a neighbor coming in, they stroll out into the fields. Let us go with them, and if their conversation is of matters pertaining to what is before them, and the scenes of their week-day labors, we think it may be none the less acceptable to Him who hath made these things for their use.

"First-rate corn weather, ain't it?"

"Splendid! Seems as if I can almost see mine grow, it pushes so fast."

"Yours all in silk?"

"Yes; and has been for a week."

"Going through it agin with the hoe?"

"No; but shall go over the field, and pull out the weeds by hand, before they go to seed."

"How do you get along haying?"

"Pretty fair; though the weather has been catching last week. I got some nice English wet in the shower."

"Have you cut that field you top-dressed last spring?"

"Yes; it's in the cock now. I've got it covered with hay caps, and shall let it sweat

to-day, only throwing the caps off in the middle of the day. I don't haul in hay Sunday, unless it looks like bad weather."

"What do you think of top-dressing?"

"Works well, what I've tried. I got at least a third more hay on that field than last year."

"S'pose 'twould pay to buy any of this patent stuff to put on an old worn-out grass field?"

"Yes. That is, if you've raked and scraped everything possible in the shape of manure you can find on or near the farm. It won't pay for us to buy these artificial manures while we let our own run to waste. That would be like trying to fill a barrel at the spigot, while the bunghole was open."

"I believe I shall try some plaster next spring."

"Ever used any on your farm?"

"Not much."

"It will probably work first-rate then; but after you use it a good many years, it don't amount to much. It's queer acting stuff too. You can't depend upon it. Some years it does first-rate, and then the next year it won't seem to be good for anything."

"How's your wheat doing?"

"Looks well, so far."

"Seen any worms in it yet?"

"No. Last year mine escaped the midge entirely, and I have hopes it will this year again. 'Twould pay pretty well again to raise it, if they would only let it alone."

"How do you s'pose they happened to let it alone last year?"

"I expect there are but few of them in the neighborhood. You know there hain't been five acres of wheat a year raised in this county for the ten years previous to the last, and so I suppose they have nearly been starved out."

"I hope they'll stay starved out."

"No such good news as that. There will probably be enough of them as soon as we get fairly at it raising wheat again."

"Potatoes are getting a pretty good growth for the season. Seen any bugs on yours yet?"

"No; it isn't quite time for them to operate much yet."

"Do you believe in hillin' 'em up so high?"

"It depends somewhat upon the soil. If it is inclined to be wet, or water stands after showers, they should be planted on top of the ground and hilled up; but if it is dry and sandy, furrow deep, and keep the field nearly level."

"I don't like the idea of hillin' corn high when you hoe it the last time."

"Nor I either, at any time. Corn roots like to run near the top of the ground, and will occupy about all the field when it gets full grown, if the ground is in such a shape that it can do so."

"Plant your peas among your potatoes yet?"

"Yes. I still stick to the old custom. I've tried every way, and I find it the best method to raise them. I know there is a good deal of talk about their injuring the potatoes, and there is some truth in it, if you plant them in the hill with them; but I put mine between the hills, covering them with the coarse dirt or sods left in covering the potatoes. It's hardly any trouble to raise them so, and you get nice, large peas."

"What kind do you raise?"

"The Marrowfats. They are large, and sell well in market."

"Some won't plant any beans or pumpkins with their corn, — say it hurts the corn crop."

"My opinion of this matter is, if your soil is rich enough, a crop of beans and pumpkins may just as well be grown with the corn; but if there is only goodness enough in the land to make a corn crop, the others had better be left out, though pumpkins do better with corn than anywhere else. Let's go over and look at my barley field."

"Seed this down this year?"

"Yes; that's one reason why I sowed the field to barley, — so as to get a good catch."

"Do you think there's much in that, after all?"

"It's the old opinion, you know, that wheat or barley is best to seed down with."

"Yes, I know; but sometimes I've had my doubts about their being any better than oats. Sow oats thin, and fix the ground as well as for barley, and grass will catch well enough. The fact is, we think a nice piece of ground is too good for oats, and only sow them on the poor, rocky, rough fields, and then find fault about the grass seed not catching well. That's been my experience."

"There is something in that. There's another thing in favor of oats, — the fodder helps pay for their cost; but barley and wheat straw is only fit for bedding. As to oats sapping the ground more than other grain, that's all humbug. Manure the oat field as well as you do for the others, and the land will be left in as good heart after the crop is taken off."

"I agree with you there."

"Let's look over the wall into the Squire's cornfield that he manured in the new way they talked so much about last year."

"New way! How was it?"

"Putting on the manure in the fall. The way they operate is to plough, harrow, furrow, and dung it out in the fall, and cover the manure just as if they had planted; then when the next spring comes, they put in the corn with a planter the very first thing, without having to wait to prepare the soil for it."

"Can get it in early — can't they?"

"Yes; that's one of the advantages claimed over the old way; and another is, that the greater part of the labor can be done in the fall when farmers are not so driven with work as in planting time."

"The Squire's corn looks pretty fair."

"I see it does. The field was in first-rate order for growing a crop. This system will operate pretty well on a field, level and gravelly, like this, but would you dare to try it on a ridge or hillside, or any inclined ground, that was in danger of washing bad by the spring rains?"

"Be likely to find your dung down in the brook, I reckon, and rather hard finding out where the hills were on the piece, too."

"That's where the danger is. One of my sowed fields washed so bad last spring that I had to plough it up and sow it over again. But there are many fields in which the method will work first rate."

"Perhaps so; tell better after trying it."

"But it's most noon; go down and take dinner with me, will you?"

"No, I thank you; mine will be ready when I get home. Good day."

www.ingramcontent.com/pod-product-compliance
Lightning Source LLC
LaVergne TN
LVHW021419110826
845150LV00007B/1990

* 9 7 8 1 4 2 5 5 0 8 8 9 0 *